USAWC STRATEGY RESEARCH PROJECT

# DEFEATING IMPROVISED EXPLOSIVE DEVICES (IED): ASYMMETRIC THREATS AND CAPABILITY GAPS

by

Colonel Gerald M. Muhl, Jr.
United States Army

Colonel Carol A. Eggert
Project Adviser

U.S. Army War College
CARLISLE BARRACKS, PENNSYLVANIA 17013

# ABSTRACT

AUTHOR: Colonel Gerald M. Muhl, Jr.

TITLE: Defeating Improvised Explosive Devices (IED): Asymmetric Threats and Capability Gaps

FORMAT: Strategy Research Project

DATE: 23 March 2011     WORD COUNT: 6,249     PAGES: 32

KEY TERMS: Explosive Ordnance Disposal (EOD), Improvised Explosive Disposal (IED), Joint IED Defeat Organization (JIEDDO), Explosive Detection Device (EDD), Exploitation, Forensics

CLASSIFICATION: Unclassified

Since our enemies are unable to defeat the U.S. Army through conventional means, they engage in sophisticated hybrid forms of warfare to exploit our vulnerabilities and threaten our national security. Because today's improvised explosive devices (IED) are not fueled by traditional military explosive materials, they avoid detection by traditional explosive detection technologies. These led to the need for military organizations and civilian agencies to research and quickly deploy new systems to counter the asymmetric IED threats. Through exploitation of IEDs, military organizations and civilian agencies can determine how they function and this knowledge will enable the Joint EOD community, and conventional troops, to defeat this threat.

# DEFEATING IMPROVISED EXPLOSIVE DEVICES (IED): ASYMMETRIC THREATS AND CAPABILITY GAPS

The term IED originated from the British Army in the 1970s, after the Provisional Irish Republican Army (IRA) used bombs made from agricultural fertilizer and Semtex (general-purpose plastic explosive similar to U.S. C4 explosive) smuggled from Libya to make highly effective booby trap devices, or bombs using command wires.[1] An IED is any device placed or fabricated in an improvised manner incorporating destructive, lethal, noxious, pyrotechnic or incendiary chemicals, designed to destroy, disfigure, distract or harass. They may incorporate military stores, but are normally devised from nonmilitary components and designed to destroy or incapacitate personnel or vehicles.[2] IEDs may incorporate military or commercially-sourced explosives and often combine both types, or they may be made with homemade explosives (HME). In short, IEDs are used to distract, disrupt, or delay an opposing force. Since most potential enemies are unable to defeat the U.S. Army through conventional means, they engage in sophisticated hybrid forms of warfare to exploit perceived vulnerabilities. Military experts anticipate hybrid tactics, such as IED's, to play a prominent role in the immediate future of warfare.[3]

IEDs have long been a challenge to the war fighter and the civilian population, but they have became a weapon of choice with the onset of Operation Iraqi Freedom (OIF) and Operation Enduring Freedom (OEF) in Afghanistan, as terrorist techniques advanced creating a truly asymmetric battlefield. As operations in Iraq and Afghanistan have continued, the IED has evolved and is becoming a serious threat to the US homeland. This asymmetric threat posed by IEDs is shaping U.S. training, tactics and strategies to prevent or at least diminish the impact of the IED. "Since the onset of the global war on terrorism in 2001, IEDs have been used extensively against coalition forces and to date they have been responsible for at least 67% of all US and coalition deaths."[4] The Department of Homeland Security and the FBI agree that the homemade explosive devices that have wreaked havoc in the battlefield pose a rising threat to the United States.[5]

IEDs are gaining traction not only worldwide but in the U.S., due in part to the relative ease of production and widespread availability of raw materials. "Because today's IEDs are not fueled by traditional military explosive materials like Trinitrotoluene (TNT), but are made in crude chemical labs using industrial chemicals like nitric acid, ammonium nitrate, diesel fuel and sugar, they avoid detection by traditional explosive detection technologies."[6] This has led to a need for military organizations and civilian agencies to research and quickly deploy new systems to counter the asymmetric IED threat.

This paper examines IEDs as an asymmetric threat to national security, identifies the shortcomings in current detection and identification capabilities, identifies the need to resource IED exploitation and offers recommendations for improving detection and identification capability and improving IED exploitation capabilities.

The IED has unfortunately made the transition from tactical "nuisance" to full-blown strategic threat, an inexpensive tactic that caused a very expensive strategic response.[7] IEDs do not go off by chance, they are an attack. The United States Secretary of Homeland Defense, Janet Napolitano stated, "A terrorist threat or incident may occur in the United States at any time and without warning and many experts believe these types of incidents can and will involve improvised explosive devices."[8] The domestic IED threat is very real. "Since 1978, the Bureau of Alcohol Tobacco and Firearms (ATF) have investigated more than 25,000 bombings and attempted bombings, more than 900 accidental explosions and more than 21,000 incidents involving recovered explosives or explosive devices. The majority of these criminal bombings involved the use of IEDs."[9] The U.S. Government will have to take this threat as seriously at home as well as abroad along with the tools and training to protect the people and the national security of the United States.

One particular example of a key vulnerability is the U.S. rail network. According to the Department of Homeland Security (DHS), terrorists' effective use of IEDs in rail attacks elsewhere in the world suggests that IEDs pose the greatest threat to U.S. rail systems. Rail systems in the U.S. have also received heightened attention as several alleged terrorists' plots have been uncovered, including multiple plots against systems in the New York City area.[10] Worldwide, terrorists have frequently targeted passenger rail systems. The most common means of attack using IEDs abroad have been on passenger trains delivered by suicide bombers. According to the Worldwide Incidents Tracking System maintained by the National Counter Terrorism Center, from January 2004 through July 2008 there were 530 terrorist attacks worldwide against passenger rail targets, resulting in more than 2,000 deaths and more than 9,000 injuries.[11] Terrorist attacks include a 2007 attack on a passenger train in India (68 fatalities and more than 13 injuries); 2005 attack on London's underground rail and bus systems (52 fatalities and more than 700 injuries); and 2004 attack on commuter rail trains in Madrid, Spain (191 fatalities and more than 1,800 injuries).[12] More recently, in January 2008,

Spanish authorities arrested 14 suspected terrorists who were allegedly connected to a plot to conduct terrorist attacks in Spain, Portugal, Germany, and the United Kingdom, including an attack on the Barcelona metro.[13] Although to date, terrorists have not yet attacked U.S. passenger rail systems, the U.S. DHS is taking proactive measures.[14]

Since October 2001, explosive devices are responsible for many of the more than 2,195 combat deaths and 21,587 wounded in action in Iraq and many of the more than 634 combat deaths and 6,046 wounded in action in Afghanistan.[15] Vehicle borne IEDs and car bombs are now used to strike police stations, markets, and mosques, killing local citizens as well as U.S. troops.[16] The victory by the US-led coalition forces in the 1991 Persian Gulf War and the 2003 invasion of Iraq, demonstrated that U.S. training, tactics and technology can provide overwhelming victories on the conventional battlefield. The Iraq campaign moved into a different type of asymmetric warfare where the coalition's use of superior conventional warfare training, tactics and technology were of much less use against continued opposition from the various partisan groups operating inside Iraq after Saddam Hussein's regime was removed from power. The IED became the weapon of choice for these groups.

A recovered IED can provide investigators a tangible item for analysis. An in depth analysis provides potential clues that help identify and understand the network responsible for the device. Explosive detection and identification is essential before any exploitation occurs. There are five primary techniques explosive ordnance disposal (EOD) and forensic investigators use to exploit IEDs. Sensitive site exploitation refers to a related series of activities taken by U.S. Government (USG) forces inside a captured sensitive site. The ability to recognize, collect, process, preserve, and analyze information, personnel, and materials found during the conduct of operations to support further analysis. Examples of this are post blast investigations, story boards, collection of forensics and biometrics at the scene that feed into the Weapons Technical Intelligence (WTI) processes. These activities exploit personnel, documents, electronic files, and material captured at the site, while neutralizing the threat posed by the site or any of its contents.[17] Biometric evidence from exploitation typically is limited to identification of bomb makers and teams that conduct IED attacks. Identification of these personnel may allow "way ahead" targeting of leadership. These are fingerprint and tissue samples taken from IED components. Mechanical and electrical exploitation of a device is the exploitation of IED components that provide information regarding the origin of the material and the logistics chain that provides the material to the bomb maker. This information helps to determine how the network operates and where the network draws support. Exploitation of explosives used in an IED provides evidence and composition; how it was put together, which leads to where the explosives came from or their logistics source of supply. Document and media exploitation (DOMEX) is advanced technology to improve the ability to organize, translate and analyze captured information in virtually all formats and many languages. The information is processed it becomes easier to find and use key data for intelligence, law enforcement and homeland defense. This means valuable resources are spent only on those documents that contain crucial clues or information.

Prior to 2001, IEDs were only rendered safe by the Joint EOD community. The Department of Defense (DoD) directed the Joint EOD Community to render safe: conventional, chemical, nuclear munitions and IEDs, support all federal government

5

agencies, support local and state government agencies, and support the U.S. Secret Service in the protection of the President of the United States and visiting foreign dignitaries. This did not allow the Joint EOD community to man, equip and train to exploit IEDs. "In February 2006 Department of Defense Directive 2000.19E further directed explosive detection, technical and forensic exploitation when exploiting IEDs."[18] This new requirement however, directed the Joint EOD community to build this capability.

Prior to September 2006, the military did not have an explosive detection capability and some would argue that we only have a partial explosive detection capability, since it only exists in Iraq and Afghanistan[19]. Taking this capability out of the lab and into the field environment takes years in the acquisition community. "The normal acquisition cycle from concept to unit fielding is six years according to Mr. Tim Walters, Senior Acquisition Analyst for the Joint Operations Support Chemical, Biological, Radiological, Nuclear and High Yield Explosives (CBRNE) Program Office."[20]

Explosives detection research crosses boundaries of physics, chemistry, materials and electronics. Trace explosive detection is a proven method for detecting explosives and is widely used as an explosives detection solution—such as in aviation security systems in most airports. "More than 7,000 explosives trace explosive detection units are currently in use by the Transportation Security Agency at U.S. airports."[21] This process involves taking a physical sample from a likely source and then analyzing it with any one of several different techniques for the presence of trace particles of explosive material.[22] Standoff explosive detection is another type of explosive detection and takes place at greater distances from people and vital assets to reduce the potential for severe damage which is why it is the most sought after capability but is still being developed and is not widely used due to technology limitations. Challenges in standoff detection include extending the distance at which effective screening can be conducted, reducing the impacts of various interferences and backgrounds (e.g., atmospheric and environmental). While it has shown some capabilities out to 1 kilometer, some of the challenges include the reliability and availability of high power, room-temperature, tunable mid-wave infrared and long-wave infrared quantum cascade lasers.[23] Another method of detection is bulk explosive detection utilizing imaging technology, much similar to the common x-ray in a hospital, to detect explosives. This capability is used to screen large objects, such as shipping containers or large pieces of luggage. The goal in bulk detection is to identify any suspicious item or anomaly which might potentially be a bomb. The equipment located in many airports throughout the world use this technology effectively.

Over the past few years, explosive detection and identification devices have become reliable and their results accepted as evidence in court cases throughout the U.S. and abroad.[24] Before September 11, 2001 and our military involvement in Iraq and Afghanistan, crime scenes and evidence processing were phenomena that typically is only observed on television or in the movies and evidence was something dealt with by the law enforcement community. Until recently, the Joint EOD community was only responsible for ensuring an IED was explosively safe and handing over what was left of the IED to the law enforcement community as evidence. Now, everything related to an IED incident is evidence, including all explosive material and test results from explosive detectors. "Evidence is defined as anything that helps us reveal proof of a fact or discover the truth of a matter, such as the identity of a person and the nature of his actions. Fortunately, wherever people go, they leave traces of themselves and take traces of their surroundings with them. As a result, criminals leave clues to their identity at crime scenes, according to Mrs. Pamela Collins, a retired U.S. Army CID agent/forensic science officer."[25] Now that the Joint EOD Community collects evidence, the U.S. Army has been tasked to provide oversight of theater evidence. The mission of overseeing theater forensic assets is new to the Army.[26] Different tactics, techniques, and procedures (TTPs) are being developed as well as closer ties with the law enforcement communities in the military, civilian and federal agencies in order to deal with this new requirement. This is another example of the military adapting to and filling the gaps in a counterinsurgency.

In the War on Terrorism, the Central Criminal Courts of Iraq (CCCI) is relying on the U.S. military to provide the evidence necessary to prosecute captured terrorists and insurgent personnel who have attacked U.S., coalition forces, Iraqi forces and Iraqi civilians.[27] The successful collection of physical evidence may mean the difference between a life sentence and the release of someone who has committed an act of terrorism. "To date, the CCCI has held 1,340 trials of insurgents suspected of anti-Iraqi and anti-Coalition activities threatening the security of Iraq and targeting Multi National Force-Iraq. These proceedings have resulted in 1,144 individual convictions with sentences ranging up to death."[28] Fortunately, it does not take years of training and field experience to be able to collect material without contaminating it; common sense and a little forethought are all that are required.[29]

Chain of custody (CoC) is also a new requirement for the Joint EOD community. Chain of custody refers to the chronological documentation or paper trail, showing the seizure, custody, control, transfer, analysis, and disposition of evidence, physical or electronic.[30] A reliable, identifiable person must always have physical custody of the evidence. In practice, this means that a police officer or law enforcement professional will take charge of a piece of evidence, document its collection, and hand it over to an evidence clerk for storage in a secure facility. CoC is also used in most explosive sampling situations to maintain the integrity of the sample by providing documentation of the control, transfer, and analysis of samples. This is especially important where explosive detection sampling can identify the existence of explosives and then be used through analysis to identify the responsible party. CoC is an important link in the process in order to prosecute either the builder of the IED, person who emplaced it or anyone that could have handled the IED from the time it was assembled to the time it was emplaced.

The ability to detect and positively identify explosives is absolutely necessary when dealing with and exploiting IEDs. From transportation to storage, ensuring the safety of personnel is critical. Risk cannot be underestimated when dealing with unknown explosives. The inventiveness and creativity of those who would do the population of the world harm is seemingly limitless. This fact has been true throughout history; today is no exception. While some people might have difficulty understanding their enemies' motivations, they can and must use their own creativity to proactively conceive adequate defenses. Getting our process "left of boom" is critical in defeating this asymmetrical threat as well as protecting our force in the exploitation cycle.[31] The following example exemplifies the need for explosive identification. On 4 November, 2007, SGT Mary Dague, a U.S. Army EOD Operator assigned to the 707th EOD Company and deployed to Iraq, lost both of her arms above the elbow transporting an unknown explosive to her vehicle.[32] This tragedy, along with many others, could have been averted if EOD could have positively identified the explosive.

The ability to provide time-sensitive, actionable intelligence to the combatant commander is the purpose of IED exploitation. The intelligence derived from forensic analysis is fused with existing intelligence regarding the insurgent or event. The result is fully integrated into existing military intelligence systems and processes and transmitted directly to the battle space owner in a timely manner so commander can maximize the use of the information. This intelligence information may also be used to prosecute insurgents through the judicial system. "Canadian Navy Petty Officer 1st Class Knobby Walsh, who recently returned from conducting counter-IED operations in Afghanistan, said the push to collect physical evidence from the bomb-making process is being welcomed by other NATO nations in the Kandahar area. EOD teams from some countries tend to destroy the devices in place, but Canadian teams prefer to disarm the bombs so they could be exploited for intelligence purposes. Walsh said that in some cases, it is possible to determine specific bomb-makers behind the devices from the tool markings on the IED. Such evidence is important in apprehending such individuals, he added. "If you don't physically give evidence to people to prosecute, then you won't be able to stop these guys," Walsh said. "And they'll just continue." He said the Canadian team was in high demand among allies in Kandahar. "They knew about the work we were doing, especially the exploitation part of it such as taking devices apart, and how important that was," Walsh said."[33]

"So how did the Joint EOD community shift to detective?  The transition came about as a result of technology that allows evidence to become a means of exposing and tracking the enemy.  EOD may not be physically present when an enemy plans and conducts an attack against friendly forces, but like the crime scene detective, EOD can examine events that have occurred and identify the enemy through the exploitation of physical evidence."[34]   The bulk of the IED exploitation work is conducted by the Combined Explosive Exploitation Cell (CEXC) with the direct support of the ATF and FBI.[35]  If further exploitation is needed, the IEDs are shipped to the Terrorist Explosive Device Analytical Center (TEDAC) in CONUS.[36]  A combined leadership team composed of the FBI and ATF manages the TEDAC and is comprised of ATF and FBI agents, intelligence analysts, certified explosives specialists, and other support personnel with specialized forensics training.  Collectively, they assist in the technical and forensic exploitation of evidence and triggering mechanisms recovered from IED detonations and render safe operations of IEDs in Iraq and Afghanistan.  The exploitation of the recovered evidence is a time consuming process in which every IED component is identified and documented in an IED database.  Additionally, these examinations identify the assembly characteristics and functionality of the IEDs.

An example of how this is put into operation is the identification of blue paint and welding techniques that played a part in this process early on in Iraq.  Shortly after a bombing, CEXC members brainstormed on how to proceed.  A CEXC member realized only a handful of welding shops in the Baghdad metropolitan areas were capable of constructing the improvised rocket launcher. While searching one of the identified

shops in the Mansur district, investigators noted that pieces of scrap metal and the shop

walls were painted the same royal blue as the generator.[37] As the investigation delved

further, the intelligence community found connections between the shop and a group of

men in Mansur already suspected in insurgent activities. "A series of coordinated raids

were successful and took members of the Al Rasheed bombing cell into custody, the

soldiers also seized cell phones, over $50,000 in cash, and computers."[38] The

combined capture of men and material proved to be a huge intelligence payoff. Equally

important, it raised awareness on the potential for another concept, linking IED

exploitation and intelligence to produce actionable intelligence. With each and every

bombing, analysis conducted led not just to the bombers, but into the whole terrorist

infrastructure. Thus, IED exploitation emerged as a focal point for thwarting the IED

insurgency in this case by tackling its asymmetric order of battle in the form of welding

techniques and blue paint. With the recent implementation of the Security Forces

Agreement between Iraq and the United States, all detentions in Iraq must now be

legally based and result from a violation of Iraqi law. An arrest warrant from an Iraqi

court is also necessary before any detention by U.S. forces. This shift from intelligence-

and security-based detentions to legal-based criminal cases was a major shift in

operations for United States Forces.[39]

The U.S. Government stood up several organizations since September 11, 2001

in order to meet the increased need for exploitation. They are a combination of U.S.

Department of Justice, U. S. Department of Defense and other agencies combining to

form a three level approach to forensic and technical exploitation of explosive devices and their respective components. Level one exploitation is accomplished by the EOD operator at the scene while either conducting a render safe operation of a device prior to detonation or during a post blast analysis after a device has detonated. Level two exploitation is forensic and technical exploitation of IEDs and is accomplished in a lab in a theater of operations. Level three exploitation is further detailed forensic and technical exploitation of IEDs in CONUS and is led by the U.S. Department of Justice. All three levels of exploitation require the capability to detect and identify explosives along with the collection of other forensics like biometrics and fingerprints. When conducting level 1-3 exploitation, there are challenges in countering the many explosives threats as there are many types and forms of explosives. The many different types of explosives are loosely categorized as military, commercial, and a third category called homemade explosives (HME) because they can be constructed with unsophisticated techniques from everyday materials. The common commercial and military explosives contain various forms of nitrogen. The presence of nitrogen is often exploited by detection technologies some of which look specifically for nitrogen (nitro or nitrate groups) in determining if a threat object is an explosive.

Military explosives include, among others, the high explosives PETN and RDX, and the plastic explosives C-4 and Semtex.[40] The military uses these materials for a variety of purposes, but mostly for demolition of unexploded ordnance, minefield clearance (minefield line clearing charge), and other specialty uses.[41] They also have commercial uses for demolition, oil well perforation, and as the explosive filler of detonation cords and explosive boosters. Military explosives cannot be purchased domestically; typically terrorists have to resort to stealing or smuggling to acquire them.

RDX was used in the Mumbai passenger rail bombings of July 2006. "PETN was used by Richard Reid, the "shoe bomber" in his 2001 attempt to blow up an aircraft over the Atlantic Ocean, and was also a component involved in the attempted bombing incident on board Northwest Airline Flight 253 over Detroit on Christmas Day 2009."[42]

Commercial explosives, with the exception of black and smokeless powders, also can only be purchased domestically by legitimate buyers through explosives distributors. Commercial explosives are often used in construction or mining activities and include, among others, trinitrotoluene (TNT), ammonium nitrate and aluminum powder, ammonium nitrate and fuel oil (ANFO), black powder, dynamite, nitroglycerin, smokeless powder, and urea nitrate.[43] Dynamite was likely used in the 2004 Madrid train station bombings, as well as the Sandy Springs, Georgia abortion clinic bombing in January, 1997. ANFO was the explosive used in the Oklahoma City, Oklahoma bombings in 1995.

In order to adapt and avoid detection, terrorists are exploring other options, such as using potassium chlorate (white, odorless powder) or sodium chlorate (yellow, odorless powder) when making IEDs. High-profile, historic examples include the London attacks in July 2005, in which suicide bombers used homemade hydrogen-peroxide based explosives to carry out attacks and the December 2001 attempted attack by "shoe-bomber" Richard Reid, who attempted to detonate a TATP explosive device while flying from Paris to Miami. Ordinary materials such as hydrogen peroxide, which is used in hair products, as a disinfectant and in swimming-pool chemicals, can be used to make bombs. Triacetone triperoxide (TATP) and hexamethylene triperoxide diamine (HMTD) are two common homemade explosives built with hydrogen peroxide.

HMEs, on the other hand, can be created using household equipment and ingredients readily available at common stores and do not necessarily contain the familiar components of conventional explosives. On February 22, 2010, Najibullah Zazi pleaded guilty to, among other things, planning to use TATP to attack the New York City subway system. Also, HMEs using TATP and concentrated hydrogen peroxide, for example, were used in the July 2005 London railway bombing. One can synthesize TATP from hydrogen peroxide and a strong acid such as sulfuric acid, and acetone, a chemical available in hardware stores and found in nail polish remover. HMTD can also be synthesized from hydrogen peroxide and a weak acid such as citric acid, and hexamine solid fuel tablets such as those used to fuel some types of camp stoves one can purchase in many outdoor recreational stores.[44]

Explosives, however, are only one component of an IED, because the various components of an IED and not just the explosive itself can also be the object of detection. Explosive systems are typically composed of a control system, a detonator, a booster, and a main charge. The control system is usually more mechanical or electrical in nature. The detonator usually contains a small quantity of a primary or extremely sensitive explosive. The booster and main charges are usually secondary explosives which will not detonate without a strong shock, for example, from a detonator. IEDs will also have some type of packaging or, in the case of suicide bombers, some type of harness or belt to attach the IED to the body. Often IEDs will also contain packs of metal such as nails, bolts, or screws or nonmetallic material which are intended to act as shrapnel or fragmentation, increasing the IED's lethality. The initiation hardware, which may be composed of wires, switches, and batteries, sets off

15

the primary charge in the detonator which, in turn, provides the shock necessary to detonate the main charge. The primary charge and the main charge are often different types and categories of explosives. "For example, in the attempted shoe bombing incident in 2001, the detonator was a common fuse and paper-wrapped TATP, while PETN was the main charge."[45] While in the past the initiation hardware of many IEDs contained power supplies, switches, and detonators, certain of the newer HMEs do not require an electrical detonator but can be initiated by an open flame.

In summary, there are numerous challenges in exploitation of IEDs to include detection of devices, identification of components to include explosives, their precursors and the forensic capability to effectively target insurgents and non state actors. These challenges prevent the Joint EOD Community from; effectively protecting the force, effectively exploiting IEDs, and the safe handling and transportation of explosives. To date, U.S. Army's effectiveness in attacking and defeating threat networks has been based upon their previous theater experience (experiential learning). The solutions to attacking networks and devices have varied by unit and have been stove-piped across the Joint EOD community as a whole. The CEXC is an ad hoc organization in Iraq that is duplicated in Afghanistan that cannot be replicated in the U.S. by any organization. The Joint EOD community is managing three different robots from two different manufacturers that use three different battery sets; just to name a few of the challenges and capability gaps.

In order to meet these challenges and close these capability gaps, the Joint Improvised Explosive Device Defeat Organization (JIEDDO) is the Department of Defense's lead counter-IED organization, dedicated to winning the fight against IEDs using all available resources.  JIEDDO was established as an organization on February

14, 2006.  JIEDDO was initially formed as the IED Task Force under the U.S. Army's

then Brigadier General Joseph Votel as an extension of their Explosive Ordnance

Disposal initiative with an obvious focus on IEDs in the fall of 2003.  The organization

was then extended to include the larger intelligence and defense communities as the

Joint IED Task Force in July 2004.  This highly classified and diverse group evolved into

the Joint IED Defeat Organization by DoD Directive 2000.19E, on February 14, 2006.[46]

Working hand-in-hand with military, government, academia, industry, and international

partners, JIEDDO is rapidly finding, developing, and delivering emerging capabilities to

counter the IED as an asymmetric weapon of strategic influence.  As part of JIEDDOs

responsibilities and functions, they rapidly acquire equipment to counter known, newly

deployed, and emerging IED threats; ensure that the systems incorporate embedded

training and logistic support; that they are fielded with a method for feedback on

effectiveness; and that they possess the flexibility for constant product improvement.

JIEDDO leads DoDs actions to rapidly provide Counter Improvised Explosive Device

(C-IED) capabilities in support of the Combatant Commanders (CCDRs) and to enable

the defeat of the IED as a weapon of strategic influence.  JIEDDO accomplishes this

along three major lines of operations:  attack the network, defeat the device, and train

the force.

In response to a Central Command (CENTCOM) Joint Urgent Operational Need

(JUON), JIEDDO worked with DoD agencies, the national test and evaluation

community, and Navy Explosive Ordnance Disposal (EOD) Technology Division

(NAVEODTECHDIV) to deliver Explosive Detection Devices (EDDs) to Iraq and

Afghanistan. The EDDs provide EOD teams the ability to rapidly identify suspicious solids, liquids, and explosives. The EDDs represent the first viable capability to identify explosives, toxic industrial chemicals (TICs), Toxic industrial material (TIMs), HME and pre-cursor components on the battlefield. Since its fielding, EOD teams are successfully using EDDs to identify HME during C-IED operations in Iraq and Afghanistan.[47] Currently, Operation New Dawn, formerly known as Operation Iraqi Freedom and Operation Enduring Freedom in Afghanistan, commercial off the shelf (COTS) EDDs are being employed to detect explosives and their precursurers. The FirstDefender and TruDefender FT are being used on a daily basis for detection and identification of both liquid and solid industrial chemicals as well as numerous explosives and their precursors. With vast databases, mixture analysis capabilities, and highly accurate, reliable operations, EDDs are being used to quickly verify the contents of tankers, drums, bags and bottles along with samples provided by EOD and Weapons Intelligence Teams (WIT) teams.[48] IED attacks are continuing at an alarming rate and the devices continue to evolve and change, therefore; technology to detect, disarm and protect war fighters must grow and improve in order to continue the fight against the asymmetric threat. Organizations, like JIEDDO, will continue to do this within the Department of Defense and others within the Department of Homeland Security and the Department of Justice will do this for the domestic threat.

The Joint EOD Community must continue its efforts to counter hybrid threats by building a technical forensics exploitation capability that provides the combatant commander these enduring tools: the use of forensic, prosecution, targeting and exploitation, in support of the interagency/joint weapons technical intelligence process.

These efforts will enhance the Joint EOD communities' ability to defeat current and emerging hybrid threats and their enabling networks since there is no question that IEDs will continue to be an asymmetric threat to national security domestically and abroad.

The next recommendation is the establishment of a Joint EOD Explosive Detection/Identification Program with the appropriate senior level program manager. This effort should be managed by the NAVEODTECHDIV, Stump Neck, Maryland. This seems a logical answer since the U.S. Navy is the DoD Executive Agent for EOD per DoD Directive 5160.62, first issued in 1971, designating the Secretary of the Navy as the Single Manager for EOD Technology and Training, a designation that continues today. This office would Identify, prioritize, and execute research and development projects that satisfy interagency requirements for existing and emerging technology in explosives detection and diagnostics. NAVEODTECHDIV could place emphasis on a long-term, sustained approach leading to new and enhanced technology for detection and identification of improvised explosive devices. NAVEODTECHDIV also could develop reliable, cost-effective advanced solutions and procedures that enhance the IED diagnostic capabilities available to the joint EOD community. Focus on the development of technologies that identify and locate the IED, explosive or enhanced fillers, and key fusing and firing components.

The next logical step is to get explosive detection into the Joint Capabilities Integration and Development System (JCIDS) process. This process encourages early and continuous collaboration with the acquisition community to ensure that new capabilities are developed for the joint EOD community. "JCIDS was created to replace the previous service-specific requirements generation system, which created redundancies in capabilities and failed to meet the combined needs of all US military services."[49] In July 2007, U.S. Defense Secretary Robert M. Gates asked Congress for approval to transfer nearly $1.2 billion to the Pentagon's Mine Resistant Ambush Protected (MRAP) program to procure an additional 2,650 vehicles. Since then, the program further evolved and is now about to include some over 15,000 vehicles. With an estimated budget of over $25 billion, MRAP is positioned to become the Defense Department's third-largest acquisition program, behind only missile defense and the Joint Strike Fighter program.[50] This demonstrates the Depart of Defense is committed to providing resources in order to protect the human capital fighting its wars; explosive detection, identification and exploitation should be no exception since it has already proven its worth by saving lives in combat.

Recommend explosive detection focus on the following lines of effort: vehicle borne improvised explosive device detection projects develop technologies to detect large quantities of explosives in vehicles at a distance. Investigate and identify the chemicals to enable detection of homemade; military; and, commercial explosives; capabilities and limitations of sensor technologies to respond to these phenomena; and enhancement of existing detection technology. Short-range detection projects develop new capabilities and improve existing systems for detection and diagnosis of terrorist devices concealed in hand baggage, cargo, and checked luggage or on persons presenting themselves at a security check point. Improve detection rate and accuracy of identifying homemade, military, and commercial explosives, as well as increase safety for both system operators and the general public. Suicide bomber detection projects develop and improve systems that detect the presence of improvised explosive devices concealed on persons engaged in suicide attacks. Systems will protect operators through standoff, where both operator and sensor are at a safe distance from the threat, or remote operation, where only the sensor is near the threat. Canine projects develop training tools, protocols, and technologies to support and enhance canine detection of explosives. Improve canine team effectiveness and consistency through better understanding of both canine detection phenomenology and canine-human interaction.[51] Specific areas of interest include equipment to enhance detection capabilities; training aids and methods to expand the number of materials detected; increased scientific understanding of canine olfactory capabilities and genetic factors favoring olfaction; innovative concepts for employing canines; and increased understanding of behavioral characteristics and rearing techniques that contribute to optimum detection performance. Attention to maintaining the health and performance of detection dogs throughout their working lives is an integral part of these efforts.

Recommend the U.S. Army move the chemical detection program, that encompasses the explosive detection program, and move it into a Joint EOD Explosive Detection/Identification Program office. "There are roughly 4,000 Ahura First Defender EDDs in use by the Joint EOD community in Iraq and Afghanistan. Evidently, there is not enough data available to make it a program of record according to Mr. Tim Walter, Senior Acquisition Analyst for the Joint Product Manager CBRNE."[52] This conflicts with the Capabilities Development for Rapid Transition (CDRT). The Army developed new materiel systems and non-materiel capabilities to meet emerging requirements to defeat the IED threat and get more "left of boom". Many of the solutions that worked well in the operational theaters have value to the Army in the long term. The goal is to significantly reduce the time needed to field selected systems or capabilities to the operational Army. The CDRT eligibility for nomination criteria requires a capability to be operationally mature, in country for a minimum of 120 days, and have a complete forward operational assessment.[53] IEDs will remain a threat in full spectrum operations. IEDs are not synonymous with or specific to the counterinsurgency environment. IEDs have a broader application to any adaptive networked threat that may challenge U.S. forces engaged across the continuum of operations, from peacetime military engagement through major combat operations. Defeating IEDs is never a standalone operation. They must be integrated into full spectrum operations in such a way that they reinforce and support the overarching campaign plan and this will only be possible if the capability gaps are filled in order to exploit IEDs to the fullest extent possible. This allows the combatant commanders the ability to fuse actionable intelligence into current operations to stay "left of boom".

This paper examined the asymmetric threat that IEDs pose to national security. The vulnerabilities of the force and the public are currently at risk due to the lack of explosive detection capability. JIEDDO is doing the heavy lifting for the current fight in Iraq and Afghanistan but the Joint EOD community needs a champion to push this effort forward into current units domestically and abroad to protect our national security. The current program needs some major improvements and the recommendations for the Joint EOD community seem in the realm of the possible even in today's resource constrained environment. Like any enduring threat, IEDs must be addressed and understood as part of our war-fighting concepts, doctrine, training, and capability development. Doing so requires a unified approach to attack the network and attacking

the device that is understood and internalized by Army organizations, Commanders,

staffs, and Soldiers, and supported by an integrated set of doctrine, organization,

training, material, leadership and education, personnel, and facilities solutions that

support these operations.

Endnotes

[1] For background on improvised explosive devices, see "Improvised Explosive Device: Background," http://en.wikipedia.org/wiki/Improvised_explosive_device (accessed February 27, 2011).

[2] U.S. Department of the Army, *Responsibilities and Procedures for Explosive Ordnance Disposal*, Army Regulation 75-15 (Washington D.C.:U.S. Department of the Army, November 1, 1978), 2.

[3] Asymmetric Warfare Public Affairs Officer, "ANT's name reflects asymmetric warfare's evolving nature," April 2003, http://www.army.mil/-news/2009/04/03/19229-ants-name-reflects-asymmetric-warfares-evolving-nature/index.html (accessed February 27, 2011).

[4] *The Joint Improvised Explosive Device Defeat Organization Home Page*, https://www.jieddo.dod.mil/about.aspx (accessed February 27, 2011).

[5] Janet Napolitano, Secretary Homeland Defense, "Keynote address to Counter IED symposium," December 1, 2009, http://www.ncsi.com/cied09/index.shtml (accessed February 27, 2011).

[6] For background on asymmetric threat, see "Explosives" located at http://www.ahurascientific.com/chemical-explosives-id/applications/explosives.php (accessed February 27, 2011).

[7] *The Joint Improvised Explosive Device Defeat Organization Home Page*, https://www.jieddo.dod.mil/about.aspx (accessed February 27, 2011).

[8] Janet Napolitano, Secretary Homeland Defense, "Keynote address to Counter IED symposium," December 1, 2009, http://www.ncsi.com/cied09/index.shtml (accessed February 27, 2011).

[9] Bureau of Alcohol Tobacco and Firearms, *Fact Sheet: EXPERTISE ON IMPROVISED EXPLOSIVES DEVICES (IEDS)*, (Washington, DC: ATF Public Affairs Division, March 2010), http://www.atf.gov/publications/factsheets/factsheet-improvised-explosives-devices.html (accessed February 27, 2011).

[10] U.S. Government Accountability Office, *TECHNOLOGY ASSESSMENT: Explosive Detection Technologies to Protect Rail Passengers: Report to Congressional* Committees, (Washington DC: U.S. Government Accountability Office, July 2010), 13.

[11] Ibid., 13.

[12] Ibid., 13.

[13] Ibid., 12.

[14] Ibid., 15.

[15] DOD Personnel and Procurement Statistics, "Personnel & Procurement Reports and Data Files," October 7, 2001 through February 7, 2011, http://siadapp.dmdc.osd.mil/personnel/CASUALTY/gwot_reason.pdf (accessed February 27, 2011).

[16] Clay Wilson, "Improvised Explosive Devices (IEDs) in Iraq and Afghanistan: Effects and Countermeasures," Updated August 28, 2007, http://www.fas.org/sgp/crs/weapons/RS22330.pdf (accessed February 27, 2011).

[17] U.S. Department of the Army, *Tactics, Techniques and Procedures for Tactical Operations involving Sensitive Sites*, Special Text 3-90.15 (Fort Leavenworth, KS: U.S. Department of the Army, Dec 2002,

[18] Department of Defense, *Directive: Joint Improvised Explosive Device Defeat Organization*, (Washington DC: U.S. Government Printing Office, February 14, 2006) http://www.dtic.mil/whs/directives/corres/pdf/200019p.pdf (accessed March 14, 2011).

[19] For information on Joint Urgent Operational Needs Statement (JUONS) submitted from MNC-I to CENTCOM for explosive detection and identification capability, see JIEDDO Home Page, *Defeat the Device: Ahura* https://www.jieddo.dod.mil/defeat.aspx (accessed February 27, 2011).

[20] Mr. Tim Walter, Senior Acquisition Analyst for the Joint Product Manager CBRNE, interviewed by author, January 17, 2011.

[21] Office of the Press Secretary, Department of Homeland Security, "Secretary Napolitano Announces Recovery Act Purchase of 1,200 Explosives Trace Detection Units to Bolster Aviation Security," April 15, 2010, http://www.hsdl.org/?view&doc=121091&coll=limited (accessed February 27, 2011).

[22] Trace particles are microscopic particles not visible to the naked eye. Existing explosives trace detectors can detect on the order of 10 nanograms of explosive trace material, which is 1,000 times smaller than what is typically considered to be the least visible amount. http://en.wikipedia.org/wiki/Explosives_trace_detector (accessed February 28, 2011).

[23] Anadi Mukherjee, Steven Von der Porten, and C. Kumar N. Patel, "Standoff detection of explosive substances at distances of up to 150 m", *Applied Optics*, Vol. 49, Issue 11, (2010):

2072-2078 http://www.opticsinfobase.org/abstract.cfm?uri=ao-49-11-2072 (accessed March 22, 2011).

[24] CEXC scientists and LE professionals testimony regarding explosive detection was accepted in an Iraqi court in 2010, see Martin Rowe, LTC, "The Forensic Exploitation Battalion", *Military Police* 19-09-1, http://www.wood.army.mil/mpbulletin/pdfs/Spring%2009/Rowe%201.pdf (accessed February 27, 2011).

[25] Pamela M. Collins, "Forensics: from its esoteric history to the streets of Baghdad", *Military Police*, PB 19-09-1, (Spring 2009): 1 http://www.wood.army.mil/mpbulletin/pdfs/Spring%2009/Collins.pdf (accessed February 28, 2011).

[26] Ibid., 2.

[27] The Central Criminal Court of Iraq, or CCCI, is a criminal court of Iraq. The CCCI is based on an inquisitorial system and consists of two chambers: an investigative court and a criminal court. The court was created by the Coalition Provisional Authority in 2003 to handle cases involving serious crimes such as governmental corruption, terrorism and organized crime that were previously handled by governorate level judges in the ordinary criminal courts. http://en.wikipedia.org/wiki/Central_Criminal_Court_of_Iraq (accessed February 28, 2011)

[28] Carolyn Anhalt, "The Iraqi Central Criminal Court convicts 16", August 6, 2006 http://www.usf-iraq.com/?option=com_content&task=view&id=1826&Itemid=21 (accessed February 28, 2011).

[29] Pamela M. Collins, "Forensics: from its esoteric history to the streets of Baghdad", *Military Police*, PB 19-09-1, (Spring 2009): 1 http://www.wood.army.mil/mpbulletin/pdfs/Spring%2009/Collins.pdf (accessed February 28, 2011).

[30] For more information on Chain of Custody http://en.wikipedia.org/wiki/Chain_of_custody (accessed 28 February, 2011).

[31] A military term for the moment before a bomb explodes, as opposed to "right of boom" which is after - based on a left to right timeline. Popularized by Washington Post writer Rick Atkinson, http://www.urbandictionary.com/define.php?term=left%20of%20boom (accessed March 23, 2011).

[32] Michelle Roberts, "War Amputee", January 9, 2009, http://thestar.com.my/lifestyle/story.asp?file=/2009/1/9/lifefocus/2873176&sec=lifefocus (accessed February 28, 2011).

[33] David Pugliese, "Canadians Launch Push Against IEDs in Afghanistan", Defense News 18 June, http://www.defensenews.com/story.php?i=3587952 (accessed February 28, 2011).

[34] Pamela M. Collins, "Forensics: from its esoteric history to the streets of Baghdad", *Military Police*, PB 19-09-1, (Spring 2009): 1 http://www.wood.army.mil/mpbulletin/pdfs/Spring%2009/Collins.pdf (accessed February 28, 2011).

[35] Beginning in 2004, the Multinational Corps–Iraq (MNC-I) established several unrelated, standalone, forensic facilities with limited mission sets. The Combined Explosive Exploitation Cell (CEXC) was the first lab established to conduct technical and limited biometric analyses on all materials related to improvised explosive devices and to develop effective countermeasures based on these analyses, see Stephen Phillips, "The Birth of the Combined Explosives Exploitation Cell" (Small Wars Journal, 2008), http://smallwarsjournal.com/blog/journal/docs-temp/52-phillips.pdf (accessed February 28, 2011).

[36] For background on TEDAC, http://www.fbi.gov/about-us/lab/tedac (accessed March 20, 2011).

[37] From combined dispatches, "U.S. arrests 35 Iraqis in hotel strike," The Washington Times, November 10, 2003 http://www. washingtontimes.com/world/20031110-120738-7694r.htm (accessed February 28, 2011).

[38] Ibid.

[39] Pamela M. Collins, "Forensics: from its esoteric history to the streets of Baghdad", Military Police, PB 19-09-1, (Spring 2009): 1 http://www.wood.army.mil/mpbulletin/pdfs/Spring%2009/Collins.pdf (accessed February 28, 2011)

[40] PETN is pentaerythritol tetranitrate. http://en.wikipedia.org/wiki/Pentaerythritol_tetranitrate; RDX is the explosive cyclotrimethylene trinitramine, also known as cyclonite. http://en.wikipedia.org/wiki/RDX; These can be used separately or combined with binders and other agents to form, for example, the hand-moldable plastic explosives, C-4http://en.wikipedia.org/wiki/C-4_(explosive) and Semtexhttp://en.wikipedia.org/wiki/Semtex. RDX is the main ingredient of C-4. Semtex contains both PETN and RDX

[41] A mine-clearing line charge (abbreviated MCLC and pronounced "mick lick") is used to create a breach in minefields under combat conditions. While there are many types, the basic design is for many explosive charges connected on a line to be projected onto the minefield. The charges explode, detonating any buried mines, thus clearing a path for infantrymen to cross. The system may either be man-portable, or vehicle-mounted. http://en.wikipedia.org/wiki/Mine-clearing_line_charge (accessed March 14, 2011).

[42] New Jersey Regional Operations Intelligence Center, Informational Advisory: Potential Tactics, Techniques and Procedures & Suspicious Activities Indicators, September 25, 2009 http://info.publicintelligence.net/MTApowderadvisory.pdf (accessed March 14, 2011).

[43] Smokeless powder is not an explosive but rather a flammable solid that burns very rapidly and is mainly used as a propellant in modern ammunitions, http://en.wikipedia.org/wiki/Smokeless_powder (accessed March 14, 2011); Black powder, also called gunpowder, is a mixture of sulfur, charcoal, and potassium nitrate. It is the main ingredient found in fireworks. In the past it was used as a propellant powder in ammunition, http://en.wikipedia.org/wiki/Black_powder, (accessed March 14, 2011).

$^{44}$ HMTD is hexamethylene tripreoxide diamine and its usual form is a white powder, http://en.wikipedia.org/wiki/Hexamethylene_triperoxide_diamine (accessed March 14, 2011); TATP is triacetone triperoxide and its usual form is a white powder http://www.globalsecurity.org/military/systems/munitions/tatp.htm (accessed March 14, 2011)..

$^{45}$ Margaret Cronin, "Judge denies bail to accused shoe bomber", CNN Online, December 28, 2001, http://articles.cnn.com/2001-12-28/us/inv.reid_1_margaret-cronin-shoe-bombs-abdul-haqq-baker?_s=PM:US (accessed March 19, 2011).

$^{46}$ Department of Defense, *Directive: Joint Improvised Explosive Device Defeat Organization*, (Washington DC: U.S. Government Printing Office, February 14, 2006) http://www.dtic.mil/whs/directives/corres/pdf/200019p.pdf (accessed March 14, 2011).

$^{47}$ *The Joint Improvised Explosive Device Defeat Organization Home Page*, https://www.jieddo.dod.mil/about.aspx (accessed February 27, 2011).

$^{48}$ Ibid.

$^{49}$ JCIDS focuses the requirements generation process on needed *capabilities* as requested or defined by one of the US combatant commanders. In the JCIDS process, regional and functional combatant commanders give feedback early in the development process to ensure that their requirements are met, Stanley A. McChrystal, *Chairman of the Joint Chiefs of Staff Instruction: Joint Capabilities Integration and Development System*, (Washington DC: CJCSSI March 1, 2009) 1.

$^{50}$ Vehicle Armoring-MRAP and Beyond, *Defense Update*, March 2007, http://defense-update.com/features/du-3-07/feature_mrap.htm, accessed March 14, 2011).

$^{51}$ Phenomenology applies to the use of sensory experiences to view and interpret, http://plato.stanford.edu/entries/phenomenology/ (accessed March 14, 2011).

$^{52}$ Mr. Tim Walter, Senior Acquisition Analyst for the Joint Product Manager CBRNE, interviewed by author, January 17, 2011.

$^{53}$ John M. McHugh and George W. Casey Jr., Americas Army, the Strength of the Nation: Army Posture Statement 2010, Posture statement presented to the 111$^{th}$ Cong, 2$^{nd}$ sess. (Washington DC: U.S. Department of the Army, 2010) https://secureweb2.hqda.pentagon.mil/vdas_armyposturestatement/2010/information_papers/Capabilities_Development_for_Rapid_Transition_(CDRT).asp (accessed March 14, 2011).